THE PHARMACY TECHNICIAN'S
Reference Guide

THE PHARMACY TECHNICIAN'S Reference Guide

Cristina Kaiser, MA

Curriculum Technician, Allied Health
San Joaquin Valley College
Visalia, California

Wolters Kluwer | Lippincott Williams & Wilkins
Health
Philadelphia · Baltimore · New York · London
Buenos Aires · Hong Kong · Sydney · Tokyo

Acquisitions Editor: David Troy
Product Manager: Andrea M. Klingler
Marketing Manager: Christen Murphy
Designer: Stephen Druding
Compositor: Macmillan Publishing Solutions

First Edition

Copyright © 2010 Lippincott Williams & Wilkins, a Wolters Kluwer business

351 West Camden Street
Baltimore, MD 21201

530 Walnut Street
Philadelphia, PA 19106

Printed in China

All rights reserved. This book is protected by copyright. No part of this book may be reproduced or transmitted in any form or by any means, including as photocopies or scanned-in or other electronic copies, or utilized by any information storage and retrieval system without written permission from the copyright owner, except for brief quotations embodied in critical articles and reviews. Materials appearing in this book prepared by individuals as part of their official duties as U.S. government employees are not covered by the above-mentioned copyright. To request permission, please contact Lippincott Williams & Wilkins at 530 Walnut Street, Philadelphia, PA 19106, via email at permissions@lww.com, or via website at lww.com (products and services).

9 8 7 6 5 4 3 2 1

Library of Congress Cataloging-in-Publication Data

Kaiser, Cristina.
  The pharmacy technician's reference guide / Cristina Kaiser.—1st ed.
      p. ; cm.
  ISBN 978-0-7817-9814-3
  1. Pharmacy technicians—Handbooks, manuals, etc. I. Title.
  [DNLM: 1. Pharmaceutical Preparations—Handbooks. 2. Pharmaceutical Preparations—administration & dosage—Handbooks. 3. Technology, Pharmaceutical—Handbooks. QV 735 K13p 2010]

RS122.95.K35 2010
615'.19—dc22

2009025320

To purchase additional copies of this book, call our customer service department at **(800) 638-3030** or fax orders to **(301) 223-2320**. International customers should call **(301) 223-2300**.

Visit Lippincott Williams & Wilkins on the Internet: http://www.lww.com. Lippincott Williams & Wilkins customer service representatives are available from 8:30 am to 6:00 pm, EST.

To daddy, for your endless support, and in memory of my mom. I love you both. Thank you for being my best friends. I have definitely been blessed to have parents like you.

DISCLAIMER

Care has been taken to confirm the accuracy of the information present and to describe generally accepted practices. However, the authors, editors, and publisher are not responsible for errors or omissions or for any consequences from application of the information in this book and make no warranty, expressed or implied, with respect to the currency, completeness, or accuracy of the contents of the publication. Application of this information in a particular situation remains the professional responsibility of the practitioner; the clinical treatments described and recommended may not be considered absolute and universal recommendations.

The authors, editors, and publisher have exerted every effort to ensure that drug selection and dosage set forth in this text are in accordance with the current recommendations and practice at the time of publication. However, in view of ongoing research, changes in government regulations, and the constant flow of information relating to drug therapy and drug reactions, the reader is urged to check the package insert for each drug for any change in indications and dosage and for added warnings and precautions. This is particularly important when the recommended agent is a new or infrequently employed drug.

Some drugs and medical devices presented in this publication have Food and Drug Administration (FDA) clearance for limited use in restricted research settings. It is the responsibility of the health care provider to ascertain the FDA status of each drug or device planned for use in their clinical practice.

Preface

The Pharmacy Technicians Reference Guide is a small, discrete guide that is easy to fit in a lab coat or smock. This reference guide is a great addition to the classroom environment; its easy access helps the student study and become acquainted with basic pharmacy information. It is also a perfect guide for the beginning Pharmacy Technician because he or she can use it to review important topics and work toward greater competency.

The information provided has been arranged for easy review and includes abbreviations, conversions, classifications, formulas, prescription information, professional titles, and terminology. Additional features include State Board websites and pharmacy organizations that are beneficial for the Pharmacy Technician. Blank sections at the bottom of pages allow space for the

Pharmacy Technician to note information relevant to the pharmacy job and help them better retain the material.

Completely searchable Full Text is available online at http://thepoint.lww.com/Kaiser. Also available there are Flash Cards, which can be used for studying and will provide help in retaining common information pertinent to Pharmacy Technician competencies.

I hope you use this guide whether you're in the classroom or any pharmacy job with confidence. Enjoy!

Contents

Classification of Drugs — 1

Anti-infectives 1
Antineoplastic 1
Biologic/Immunologic 2
Blood/Vessels 2
Cardiovascular 3
Central Nervous System 4
Dermatologic 6
Endocrine 6
Excretory/Renal 7
Gastrointestinal 7
Ophthalmic 8
Otic 8
Respiratory 8

Abbreviations — 10

Abbreviations Not Accepted by JCAHO 10
Dosage Forms 10
Drugs 12
Drug Forms 16
Measurements 17
Medical Terminology 19
   Routes of Administration 31
   Units of Time 33
   Vitamins 35
   Diseases 36

Formulas　　　　　　　　　　　　　　　　41

Dosage Calculations　41
　Body Surface Area　41
　Volume　43
　Percent of a Quantity　44
　Ratio–Proportion Method　44
　Solutions　44

FDA-Mandated Medication Guides　45

Hospital Chart Information　46

Order Requirements　46
Maximum Daily Dosages　47

Metric Conversions　48

Household Metric Conversions　48
　Volume　48
　Weight　49
　Dry Measure　49

Prefixes and Suffixes　50

Prefixes　50
Suffixes　54

Pregnancy Categories　57

Professional Titles　59

Retail Hardcopy Information　63

Patient Information　63
Expiration Date of Prescriptions　63

CONTENTS

Roman Numerals — 64

Inappropriate Medications for Seniors — 68

Antianxiety Agents 68
Antiemetic Agents 68
Antidepressant Agents 68
Antihypertensive Agents 69
Narcotic Analgesics 69
Sedative–Hypnotic Agents 70
Sulfonylurea 70

Temperature Scales — 71

24-Hour Clock — 72

Pharmacy Organizations — 75

State Boards of Pharmacy — 76

Classifications of Drugs

Anti-infectives

Amebicide
Antifungal
Antimalarial
Antiretroviral
  Fusion inhibitor
  Nucleoside analog
    reverse
    transcriptase
  Nucleoside reverse
    transcriptase
  Protease inhibitor
Antiviral
Bacitracin
Cephalosporine
Fluoroquinolone
Lincosamide
Lipopeptide
Macrolide
Methenamine
Quinolone
Sulfonamide
Tetracycline

Antineoplastic

Alkylating
  Alkyl sulfonate
  Nitrogen mustard
  Nitrosourea
  Triazene
Antimetabolites
Antimitotic
  Taxoid
  Vinca alkaloid

CLASSIFICATIONS OF DRUGS

Chemotherapy regimen
DNA
Enzyme
Hormone
  Androgen
  Antiandrogen
  Antiestrogen
  Aromatase
  Progestin
Proteasome inhibitor
Protein-tyrosine kinase inhibitor
Retinoide
Rexinoid

Biologic/Immunologic

Active Immunization
  Bacterial vaccine
  Toxoid
  Viral vaccine
Antirheumatic
  Gold compounds
Antitoxin/antivenin
Immune globulin
Immunologic
  Immunomodulator
  Immunostimulant
  Immunosuppressive

Blood/Vessels

Anticoagulant
  Antithrombin
  Heparin
  Low molecular weight heparins
  Selector factor Xa inhibitor
  Thrombin inhibitor
  Warfarin
Antiplatelet
  Aggregation inhibitor
  Antiplatelet combination

Glycoprotein IIb/IIIa
Coagulant
  Heparin antagonist
Plasma expanders
  Dextran adjunct
  Plasma protein fractions
Thrombolytic
  Recombinant human activated protein C
  Thrombolytic enzymes
  Tissue plasminogen activators

Cardiovascular

Antiadrenergic/sympatholytic
  Alpha/beta-adrenergic blocker
  Antiadrenergic
    Centrally acting
    Peripherally acting
  Beta-adrenergic blocker
Antiarrhythmic
Antihyperlipidemic
  Bile acid sequestrant
  Fibric acid derivative
  HMG-CoA reductase inhibitor
Calcium channel blocker
Renin angiotensin system antagonist
  Angiotensin-converting enzyme inhibitor
  Angiotensin II receptor antagonist
  Selective aldosterone receptor antagonist

CLASSIFICATIONS OF DRUGS

Vasodilator
  Endothelin receptor antagonist
  Human B-type natriuretic peptide
  Nitrate
  Peripheral vasodilator
  Prostacyclin analog

Central Nervous System

Anesthetic
Antianxiety
  Benzodiazepine
Anticonvulsant
  Benzodiazepine
  Hydantoin
  Succinimide
  Sulfonamide
Antidepressant
  Monoamine oxidase inhibitor (MAOI)
  Selective serotonin reuptake inhibitor (SSRI)
  Serotonin and norepinephrine reuptake inhibitor
  Tetracyclic compound
  Tricyclic compound
Antiemetic/antivertigo
  5-HT$_3$ receptor antagonist
  Anticholinergic
  Antidopaminergic
Anti-inflammatory
  Nonsteroidal
    Selective COX-2 inhibitor
Antiparkinson
  Anticholinergic

Dopaminergic
Antipsychotic
  Benzisoxazole derivative
  Dibenzapine derivative
  Dihydroindolone derivative
  Phenothiazine derivative
  Phenylbutylpiperadine derivative
  Quinolinone derivative
  Thioxanthene derivative
CNS stimulant
  Analeptic
  Amphetamine
  Anorexiant
Migraine
  Ergotamine derivative
  Serotonin 5-HT$_1$ receptor agonist
Muscle relaxant
  Depolarizing neuromuscular blocker
  Nondepolarizing neuromuscular blocker
  Skeletal
    Centrally acting
    Direct acting
Opioid analgesic
Salicylate
Sedative and hypnotic
  Nonbarbiturate
  Barbiturate
    Intermediate acting
    Long acting
    Short acting
Smoking deterrent

CLASSIFICATIONS OF DRUGS

CLASSIFICATIONS OF DRUGS

Dermatologic

Anti-infective
Anti-inflammatory
Astringent
Cleanser
Diaper rash
Keratolytic
Sunscreen
Photochemotherapy
Pigment agent
Retinoid
Scabicide/pediculicide
Wound healing agent

Endocrine

Adrenocortical steroid
  Adrenal steroid inhibitor
  Glucocorticoid
  Mineralocorticoid
Antidiabetic
  Alpha-glucosidase inhibitor
  Amylin analog
  Antidiabetic combination
  Biguanide
  Incretin mimetic
  Insulin
  Meglitinide
  Sulfonylureas
  Thiazolidinedione
Detoxification
  Antidote
  Chelating agent
Sex hormone
  Anabolic steroid
  Androgen
  Androgen hormone inhibitor
  Contraceptive hormone
  Estrogen
  Estrogen and progestin combination
  Ovulation stimulant

Progestin
Selective estrogen
  receptor modulator
Thyroid
  Antithyroid agent
  Thyroid hormone

Excretory/Renal

Anticholinergic
Diuretic
  Carbonic anhydrase
    inhibitor
  Diuretic
    combination
  Loop diuretic
  Nonprescription
    diuretic
  Osmotic diuretic
  Potassium-sparing
    diuretic
  Thiazide and related
    diuretic
Genitourinary irrigant
  Hexitol
  Neomycin and
    polymyxin B
Impotence
  Phosphodiesterase
    type 5 inhibitor
Urinary acidifier
  Acid phosphate
  Ascorbic acid
Urinary alkalinizer
Vaginal preparation
  Anti-infective
    miscellaneous
  Vaginal antifungal
    agent

Gastrointestinal

Antacid
Antidiarrheal
Antiflatulent
Gastrointestinal
  anticholinergics/
  antispasmodic

CLASSIFICATIONS OF DRUGS

Histamine H_2 antagonist
Gastrointestinal stimulant
Laxative
Lipase inhibitor
Proton pump inhibitor
Prostaglandin
Sucralfate

Ophthalmic

Antibiotic
Sulfonamide
Antiseptic
Artificial tear
Corticosteroid
Glaucoma
  Alpha-adrenergic agonist
  Alpha-adrenergic antagonist
  Beta-adrenergic blocking agent
  Carbonic anhydrase
  Miotics
    Cholinesterase
    Direct acting
  Prostaglandin agonist
  Sympathomimetic
Ocular lubricant
Punctal plug
Surgical adjunct

Otic

Otic preparation

Respiratory

Antiasthmatic combination
  Xanthine
  Xanthine-sympathomimetic
Antihistamine
  Alkylamine

Ethanolamine
Phenothiazine
Phthalazinone
Piperazine
  Nonselective
  Peripherally selective
Piperidine
  Nonselective
  Peripherally selective
Bronchodilator
  Anticholinergic
  Diluents
  Sympathomimetic
  Xanthine derivative
Expectorant
Nasal decongestant
  Arylalkylamine
  Imidazoline
Nonnarcotic antitussive
Respiratory inhalant
  Corticosteroid
  Intranasal steroid
  Mast cell stabilizer
  Mucolytic
  Respiratory gases
Upper respiratory combination

Notes: _____
_____
_____
_____
_____
_____
_____

CLASSIFICATIONS OF DRUGS

Abbreviations

Abbreviations Not Accepted by JCAHO

1.0	MS
.1	MSO_4
CC	QD
U	QID
IU	QOD
$MGSO_4$	TIW

For further information, visit:
http://www.jointcommission.org/PatientSafety/DoNotUseList

Dosage Forms

AQ; H_2O	Water
Aqua Dist	Distilled water
Caps	Capsules

(continued)

Dosage Forms *(continued)*

DPI	Dry powder inhaler
DW	Distilled water
D_5W	Dextrose 5% in water
ELIX	Elixir
GTT	Drop
LIQ	Liquid
MDI	Metered dose inhaler
NS	Normal saline
SOLN; SOL	Solution
SUPP	Suppository
SUSP	Suspension
SYR	Syrup
TAB	Tablet

(continued)

Dosage Forms *(continued)*

TR; TINCT	Tincture
U	Unit
UD	Unit dose
UNG; μ	Ointment

Drugs

APAP	Acetaminophen
ASA	Aspirin
ASP	Aspartic acid
BZD	Benzodiazepine
Ca	Calcium
Cl	Chloride
CO_2	Carbon dioxide
COD	Codeine

(continued)

Drugs *(continued)*

Cu	Copper
F	Fluoride
Fe	Iron
$FeSO_4$	Ferrous sulfate
FRU	Fructose
GLY	Glycine
H2	Histamine
H_2O	Water
HC	Hydrocortisone
HCl	Hydrochloric acid
HCTZ	Hydrochlorothiazide
Hg	Mercury
I	Iodine

(continued)

Drugs *(continued)*

INH	Isoniazide
K	Potassium
LYS	Lysine
Mg	Magnesium
MPA	Medroxyprogesterone
MS; MSO4	Morphine sulfate
MTX	Methotrexate
Na	Sodium
NTG	Nitroglycerine
O	Oxygen
P	Phosphorus
PABA	Para-aminobenzoic acid
PAS	Para-aminosalicylic acid

(continued)

Drugs *(continued)*

PEG	Polyethylene glycol
Phe	Phenylalanine
Se	Selenium
SMZ	Sulfamethoxazole
T3	Tylenol with codeine 300 mg/30 mg
T3	Triiodothyronine
T4	Thyroxine
TAC	Triamcinolone
TCN	Tetracycline
TMP	Trimethoprim
Zn	Zinc

Drug Forms

AQ	Water base
CD	Control delay
DS	Double strength
EC	Enteric coated
ER	Extended release
HFA	Contains no chlorofluorocarbons
LA	Long acting
ODT	Orally dissolving tablet
SA	Sustained acting
SF	Sugar free
SR	Sustained release
T½	Half life
TD	Time delay
TR	Time release
XL; XR; XT	Extended release

Measurements

AA	Of each
AD	To; up to
AQ; AD	Add water up to
C; °C	Celsius
CM	Centimeter
DIL	Dilute
DIV	Divide
F; °F	Fahrenheit
F; FL	Fluid
G, g; GM, gm	Gram
Gr	Grain
GTTS	Drops
GTT	Drop
HT, ht	Height

(continued)

ABBREVIATIONS | **MEASUREMENTS** 18

Measurements *(continued)*

Kg	Kilogram
L	Liter
LB, lb	Pound
m²	Square meter
MCG	Microgram
mEq	Milliequivalent
MG, mg	Milligram
ml, mL	Milliliter
mm	Millimeter
No; #	Number
QS	Sufficient quantity
QS AD	Add sufficient quantity to make
QTY	Quantity

(continued)

Measurements *(continued)*

T	Temperature
TBSP; TBL	Tablespoonful
TSP; TEA	Teaspoonful
U	Unit
WT	Weight

Medical Terminology

AD	Up to
AGIT	Agitation
AMI	Acute myocardial infarction
ANS	Autonomic nervous system
AQ; H_2O	Water

(continued)

ABBREVIATIONS | MEDICAL TERMINOLOGY

Medical Terminology *(continued)*

BE	Barium enema
BM	Bowel movement
BP	Blood pressure
BPM	Beats per minute
BS	Blood sugar
BSA	Body surface area
BX	Biopsy
Δ	Change
CA	Cancer
CAD	Coronary artery disease
CAT	Computed axial tomography
CATH	Catheter
CCU	Coronary care unit

(continued)

Medical Terminology *(continued)*

CDS	Controlled drug substance
CHD	Coronary heart disease
CHF	Congestive heart failure
CNS	Central nervous system
COMP; CPD	Compound
CSF	Cerebrospinal fluid
CT	Computed tomography
CV	Cardiovascular
CX	Cervix
D&C	Dilation and curettage
DNR	Do not resuscitate
DOB	Date of birth
DTP	Diphtheria, tetanus toxoids, and pertussis vaccine

(continued)

Medical Terminology *(continued)*

DUE	Drug usage evaluations
DUR	Drug utilization review
DVT	Deep venous thrombosis
DX	Diagnosis
ECG; EKG	Electrocardiogram
ECU	Emergency care unit
EEG	Electroencephalogram
ENT	Ear, nose, and throat
ER	Emergency room
EUS	Endoscopic ultrasound
FAS	Fetal alcohol syndrome
FX	Fracture
GI; GIT	Gastrointestinal tract
HA	Headache

(continued)

Medical Terminology *(continued)*

Hb; Hbg	Hemoglobin
HBP	High blood pressure
HCG	Human chorionic gonadotropin
HD	Hemodialysis
HDL	High-density lipoprotein
HEPA	High-efficiency particulate air
HIPAA	Health Insurance Portability and Accountability Act
HIV	Human immunodeficiency virus
HR	Heart rate
HX	History

(continued)

Medical Terminology *(continued)*

IC	Intracoronary
ICF	Intracellular fluid
ICP	Intracellular pressure
ICU	Intensive care unit
IP	Inpatient
IVP	Intravenous pyelogram
LDL	Low-density lipoprotein
LTM	Long-term memory
MAO	Monoamine oxidase
MAOI	Monoamine oxidase inhibitor
MAR	Medication administration record
MI	Myocardial infarction

(continued)

Medical Terminology *(continued)*

MIP	Maximum inspiratory pressure
MMR	Measles, mumps, and rubella virus vaccine (live)
Mol	Mole
MRI	Magnetic resonance imaging
NKA	No known allergies
NKHC	No known health conditions
N/V	Nausea and vomiting
N/V/D	Nausea, vomiting, and diarrhea
OD	Overdose
OP	Outpatient

(continued)

Medical Terminology *(continued)*

OR	Operating room
P	Pain
PAR	Postanesthetic recovery
PE	Pulmonary embolism
PEDS	Pediatrics
PMR	Patient medication record
PMS	Premenstrual syndrome
POS	Point of service
POST OP	After surgery
PPI	Patient package insert
PRE OP	Before surgery
PSA	Prostate-specific antigen
PT	Patient; physical therapy

(continued)

Medical Terminology *(continued)*

RBC	Red blood cell; red blood count
RDS	Respiratory distress syndrome
REM	Rapid eye movement
Rh	Rhesus
rpm	Revolutions per minute
rps	Revolutions per second
RR	Respiratory rate
Rx	Prescription; take
SOB	Shortness of breath
SSRI	Selective serotonin reuptake inhibitors
STAPH.	*Staphylococcus*
STM	Short-term memory

(continued)

Medical Terminology *(continued)*

SX	Symptoms
TBW	Total body weight
TCA	Tricyclic antidepressant
TIA	Transient ischemic attack
TLC	Total lung capacity
TMJ	Temporomandibular joint
TNM	Tumor, node, metastasis
TOPV	Trivalent oral polio vaccine
TPN	Total parenteral nutrition
TPR	Temperature, pulse, respirations; third party reject
TQM	Total quality management
TSA	Tumor-specific antigens

(continued)

Medical Terminology *(continued)*

TSH	Thyroid-stimulating hormone
TSS	Toxic shock syndrome
TSTA	Tumor-specific transplantation antigen
TV	Tidal volume
TX	Treatment
U	Unit
UA	Urinalysis
UD	Unit dose
URI	Upper respiratory infection
UTI	Urinary tract infection
VC	Vital capacity

(continued)

Medical Terminology *(continued)*

VHDL	Very high density lipoprotein
VLDL	Very low density lipoprotein
VO	Verbal order
VS	Vital signs
V/V	Volume-to-volume ratio
V/W	Volume-to-weight ratio
WBC	White blood cell count
WT	Weight
W/V	Weight-to-volume ratio
W/W	Weight-to-weight ratio
XX	Female sex chromosome
XY	Male sex chromosome
YO	Years old

Routes of Administration

A.D., AD	Right ear
A.S., AS	Left ear
AU	Both ears; each ear
BUC	Buccal
ID	Intradermal
IM	Intramuscular
INJ	Injection
IP	Intraperitoneal
IT	Intrathecal
IV	Intravenous
IVP	Intravenous push
IVPB	Intravenous piggyback
L; l	Left
LVP	Large-volume parenterals

(continued)

Routes of Administration *(continued)*

NPO	Nothing by mouth
OD	Right eye
OPTH	Eye
OS	Left eye
OTIC	Ear
OU	Both eyes; each eye
PER	By; through
PO	By mouth; orally
PR	Per rectum/rectally
PV	Per vagina/vaginally
R, r	Right
SL	Sublingual
SQ; SC	Subcutaneous
TOP	Topically

Units of Time

A	Before
AC	Before meals
AM	Morning
ATC	Around the clock
BID	Twice a day
BIW	Twice a week
C, c; c̄	With
D, d	Day; daily
D/C	Discontinue
FID	Five times a day
H; °	Hour
HS	At bedtime; hour of sleep
MO	Month
NOC; N, n	Night

(continued)

Units of Time *(continued)*

P	After
PC	After meals
PM	After noon; evening
PRF	As needed for
PRN	As needed
Q	Every
QD	Every day
QH	Every hour
QHS	Every night at bedtime
QID	Four times a day
QOD; QAD	Every other day
QV	As much as you wish
STAT	Immediately
TID	Three times a day

(continued)

Units of Time *(continued)*

TIW	Three times a week
WA	While awake
WK	Week
X, x	Times; for
YR	Year

Vitamins

A	Retinol
B_1	Thiamin
B_2	Riboflavin
B_3	Niacin
B_6	Pyridoxine
B_7; H	Biotin
B_9	Folic acid

(continued)

Vitamins (continued)

B_{12}	Cyanocobalamin
C	Ascorbic acid
D	Calciferol
E	Tocopherol
K_1	Phylloquinone
K_2	Menaquinone
K_3	Menadione

Diseases

AD	Alzheimer's disease
ADD	Attention deficit disorder
ADHD	Attention deficit hyperactivity disorder
AIDS	Acquired immune deficiency syndrome

(continued)

Diseases *(continued)*

ALS	Amyotrophic lateral sclerosis
AMI	Acute myocardial infarction
CA	Cancer
CAD	Coronary artery disease
CF	Cystic fibrosis
CHD	Coronary heart disease
CHF	Congestive heart failure
COPD	Chronic obstructive pulmonary disorder
CP	Cerebral palsy
CRD	Chronic respiratory disease

(continued)

ABBREVIATIONS | MEDICAL TERMINOLOGY

Diseases *(continued)*

CTS	Carpal tunnel syndrome
DJD	Degenerative joint disease
DVT	Deep venous thrombosis
EBV	Epstein-Barr virus
ED	Erectile dysfunction
FAS	Fetal alcohol syndrome
GERD	Gastroesophageal reflux disease
HA	Headache
HPV	Human papillomavirus
HSV	Herpes simplex virus
IBS	Irritable bowel syndrome
IH	Infectious hepatitis

(continued)

Diseases (continued)

LDL	Low-density lipoprotein
MI	Myocardial infarction
MS	Multiple sclerosis; mitral stenosis
PE	Pulmonary embolism
PMS	Premenstrual syndrome
PUD	Peptic ulcer disease
RA	Rheumatoid arthritis
RDS	Respiratory distress syndrome
SIDS	Sudden infant death syndrome
SOB	Shortness of breath
STD	Sexually transmitted disease

(continued)

Diseases *(continued)*

STR	*Streptococcus*
TB	Tuberculosis
TIA	Transient ischemic attack
TSS	Toxic shock syndrome
URI	Upper respiratory infection
UTI	Urinary tract infection
VD	Venereal disease

Notes: _____

FORMULAS

Dosage Calculations

Body Surface Area

$$\text{Metric BSA (m}^2\text{)} = \frac{D\,(\text{desired})}{H\,(\text{on hand})} \times Q\,(\text{quantity})$$
$$= X\,(\text{amount})$$

Conversion Factor Method

$$\frac{D}{M} \times E = Q$$

IV Flow Rate

$$\frac{\text{Total mL}}{\text{Total h}} = \text{mL/h}$$

$$\frac{\text{Total mL}}{\text{Total min}} \times 60\,\text{min/h} = \text{mL/h}$$

FORMULAS | DOSAGE CALCULATIONS

$$\frac{V(\text{mL})}{T(\text{min})} \times C\,(\text{gtt/mL}) = R\,(\text{gtt/min})$$

IV Flow Rate for Electronic Regulators

$$\frac{\text{Total mL ordered}}{\text{Total h ordered}} = \text{mL/h}$$

Remember to round mL/h to a whole number.

IV Flow Rate for Manually Regulated IVs

$$\frac{V(\text{vol})\,\text{mL}}{T(\text{time in min})} \times C\,(\text{calibration or drop factor, gtt/mL}) = R\,(\text{rate})$$

IV Flow Rate Including IV PB

$$\text{Rate:}\ \frac{V}{T} \times C = R$$

or

$$\frac{\text{mL/h}}{\text{Drop factor constant}} = R$$

Time: Time for 1 dose × number of doses within 24 hours

Volume

IV Flow Rate (Recalculation of Off-Schedule IV Flow Rate)

1. $\dfrac{\text{Remaining volume}}{\text{Remaining hours}}$ = Recalculated mL/h

2. $\dfrac{V}{T} \times C$ = gtt/min

3. $\dfrac{\text{Adjusted gtt/min} - \text{Ordered gtt/min}}{\text{Ordered gtt/min}}$
    = % Variation

IV Flow Rate Shortcut

$$\dfrac{\text{mL/h}}{\text{Drop factor constant}} = R \text{ (gtt/min)}$$

IV Infusion Time

$$\dfrac{\text{Total volume}}{\text{mL/h}} = \text{Total hours or mL/h} = \dfrac{\text{Total mL}}{\times \text{ Total h}}$$

IV Volume

Total hours × mL/h = Total volume

FORMULAS | DOSAGE CALCULATIONS

FORMULAS | DOSAGE CALCULATIONS

Percent of a Quantity

Percentage (part) = Percent × whole quantity

Ratio–Proportion Method

$$\frac{\text{Dosage on hand}}{\text{Volume on hand}} = \frac{\text{Dosage desired}}{\times \text{Volume}}$$

Ratio–Proportion Method (Two Equivalent Ratios)

$$\frac{A}{B} = \frac{C}{D}$$

or A : B = C : D

or A : B :: C : D

Solutions

1. *D* (Desired solution strength) × *Q* (Quantity of desired solution) = × (Amount of solute)
2. Quantity of desired solution − Amount of solute = Amount of solvent

Notes: _____
_____
_____
_____
_____
_____

FDA-Mandated Medication Guides

- Ambien CR
- Antidepressants
- Avonex
- Elidel
- Emsam
- Epzicom
- EXUBERA
- Foradil
- Forteo
- INFERGEN
- Isotretinoin
- Lariam
- Lindane
- Lotronex
- Mifeprex
- Nolvadex
- Nonsteroidal anti-inflammatory drugs
- Pacerone
- PEG-Intron
- PEGASYS
- Protopic
- REBETRON
- Rebif
- REVLIMID
- Ribavirin
- Roferon A
- Serevent DISKUS
- Soriatane
- Strattera
- SYMLIN
- Tracleer
- Trizivir
- Tysabri
- Viramune
- Xyrem
- Ziagen

HOSPITAL CHART INFORMATION

Order Requirements

Patient name/gender
Weight or BSA
Medical record number
Account number
Allergies
Allergies to medications
Drug name and strength
Date/time
Physician's signature

Maximum Daily Dosages

Acetaminophen	4 gm
Aspirin	4 gm
Calcium	1200 mg
Cephalexin	4 gm
Gabapentin	1800 mg
Ibuprofen	32 gm
Megestrol acetate	80 mg
Tussionex	10 mL/24 h

Notes: _____
_____
_____
_____
_____
_____

Metric Conversions

1 gram (gm) = 1000 mg
1 milligram (mg) = 1000 mcg (µg) = 0.001 gm
1 microgram (mcg) = 0.001 mg = 0.000001 gm
1 kilogram (kg) = 1000 gm
1 liter (L) = 1000 mL
1 milliliter (mL) = 0.001 L
1 meter (m) = 100 cm = 1000 mm
1 centimeter (cm) = 0.01 m = 10 mm
1 millimeter (mm) = 0.001 m = 0.1 cm

Household Metric Conversions

Volume

1 cc = 1 mL (or ml)
5 mL = 1 teaspoon (tea/tsp)
15 mL = 3 tea = 1 tablespoon (tbl)
= 1/2 ounce (oz)
30 mL = 6 tea = 2 tbl = 1 oz

240 mL = 1 cup = 1/2 pint (pt)
480 mL = 16 oz = 2 cups = 1 pt
960 mL = 32 oz = 2 pt = 1 quart (qt)
1000 mL = 1 Liter (L)
3840 mL = 8 pt = 4 qt = 1 gallon (G)

Weight

1 mg = 1000 mcg
60 mg/65 mg = 1 grain (gr)
300 mg/325 mg = 5 gr
1 dram = 60 gr
1 gm = 1000 mg
30 gm/28.35 gm = 1 oz
454 gm = 1 lb
1 kg = 1000 gm
2.2 lb = 1 kg

Dry Measure

1 lb = 12 oz = 373.2 gm

Notes: _____
_____
_____

PREFIXES AND SUFFIXES

Prefixes

a-	Without
ambi-	Both
ante-	Before
anti-	Against
bi-	Two
brady-	Slow
cirrh-	Yellow
con-	With
contra-	Against
de-	Reduce
dia-	Across; through

(continued)

Prefixes *(continued)*

dis-	Separate from; apart
dys-	Painful; difficult
ecto-	Outside
end-	Within
epi-	Upon
eu-	Good; normal
exo-	Outside
heter-	Different
hom-	Same
hyper-	Above; in excess
Hypo-	Below; lacking
immun-	Safe; protected
infra-	Below or under

(continued)

Prefixes *(continued)*

inter-	Between
intra-	Within
leuk-	White
macro-	Large
medi-	Middle
meso-	Middle
meta-	Beyond; after; changing
micro-	Small
mono-	One
neo-	New
ot-	Ear
para-	Alongside; abnormal
peri-	Around
post-	After
pre-; pro-	Before

(continued)

Prefixes *(continued)*

pseudo-	False
quadri-	Four
re-	Again; back
semi-	Half
sub-	Under
super-; supra-	Greater than
sym-; syn-	With
tachy-	Fast
trans-	Across; through
tri-	Three
ultra-	Beyond
uni-	One
xer-	Dry

Suffixes

-ac; -al; -ar; -ary	Relating to
-algia; -cynia	Pain
-asthenia	Not having strength
-cele	Pouching; hernia
-cyesis	Pregnancy
-eal	Pertains to
-ectasis	Dilation
-ectomy	Removal
-emia	Blood condition
-gram	Record
-graphy	Recording process
-ia; -ism	Condition of
-iasis	Formation of
-iatry	Treatment

(continued)

Suffixes *(continued)*

-ic	Pertains to
-itis	Inflammation
-ium	Tissue
-logy	Study of
-malacia	Softens
-megaly	Enlargement
-ole	Small
-oma	Tumor
-osis	Abnormal condition
-paresis	Partial paralysis
-pathy	Disease
-penia	Reduce
-phagia	Swallowing
-phasia	Speech
-philia	Attraction for

(continued)

PREFIXES AND SUFFIXES | SUFFIXES

Suffixes *(continued)*

-phobia	Fear of
-plasia	Formation of
-rrhea	Discharge
-sclerosis	Constriction
-scopy	Examination
-stasis	Stop; stand
-tomy	Incision
-toxic	Poison
-tropic	Stimulation
-ula	Small

Notes: _____
_____
_____
_____
_____
_____

Pregnancy Categories*

Category A: *Controlled studies show no risk.* Adequate and well-controlled studies have failed to demonstrate a risk to the fetus in the first trimester of pregnancy (and there is no evidence of risk in later trimesters).

Category B: *No evidence of risk in humans.* Animal reproduction studies have failed to demonstrate a risk to the fetus and there are no adequate and well-controlled studies in pregnant women.

Category C: *Risk cannot be ruled out.* Animal reproduction studies have shown an adverse effect on the fetus and there are no adequate and well-controlled studies in humans, but potential

*Adapted from the FDA Pregnancy Categories. Available at: http://depts.washington.edu/druginfo/Formulary/Pregnancy.pdf.

benefits may warrant use of the drug in pregnant women despite potential risks.

Category D: *Positive evidence of risk.* There is positive evidence of human fetal risk based on adverse reaction data from investigational or marketing experience or studies in humans, but potential benefits may warrant use of the drug in pregnant women despite potential risks.

Category X: *Contraindicated in pregnancy.* Studies in animals or humans have demonstrated fetal abnormalities and/or there is positive evidence of human fetal risk based on adverse reaction data from investigational or marketing experience, and the risks involved in use of the drug in pregnant women clearly outweigh potential benefits.

Notes: _____
_____
_____
_____
_____
_____

Professional Titles

AP	Acupuncture physician
CMA	Certified medical assistant
CMT	Certified medical transcriptionist
CNA	Certified nursing assistant
CNM	Certified nurse midwife
CPhT	Certified pharmacy technician
CRNA	Certified registered nurse anesthetist
CRNP	Certified registered nurse practitioner

(continued)

PROFESSIONAL TITLES

DC	Doctor of chiropractic
DDS	Doctor of dental surgery
DMD	Doctor of dental medicine
DNS	Director of nursing services; doctor of nursing services
DO	Doctor of osteopathy
DP	Doctor of podiatry
DPH	Doctor of public health; doctor of public hygiene
DPM	Doctor of physical medicine; doctor of podiatric medicine
DVM	Doctor of veterinary medicine
FNP	Family nurse practitioner
LN	Licensed nutritionist

(continued)

LPN	Licensed practical nurse
LVN	Licensed visiting nurse; licensed vocational nurse
MD	Doctor of medicine
MNNP	Master of nursing; nurse practitioner
MSN	Master of nursing
ND	Doctor of neuropathic medicine
NP	Nurse practitioner
OD	Doctor of optometry
PA	Physician assistant
PA-C	Physician assistant certified
PharmD	Doctor of pharmacy
PhD	Doctor of philosophy

(continued)

PROFESSIONAL TITLES

PsyD	Doctor of psychology
PhT	Pharmacy technician
PT	Physical therapist
RN	Registered nurse
RN-C	Registered nurse certified
RNCS	Registered nurse clinical specialist
RN/NP	Registered nurse/nurse practitioner
RPh	Registered pharmacist
VMD	Veterinary medical doctor

Notes: _____

Retail Hardcopy Information

Patient Information

Name _____

Address _____

Phone number _____

Date of birth _____

Allergies to medications _____

Health conditions _____

Sex (male/female) _____

Private pay or insurance _____

Expiration Date of Prescriptions

OTC/Noncontrols: 1 year
Controls 2–5: 6 months

Roman Numerals

Remember: The smaller numeral is subtracted from the larger numeral when the smaller numeral is before a larger numeral. Add the numerals when the smaller numeral follows the larger.

Arabic Numeral	Small Roman Numeral	Large Roman Numeral
½	ss	SS
1	i	I
2	ii	II
3	iii	III
4	iv	IV

(continued)

Roman Numerals *(continued)*

Arabic Numeral	Small Roman Numeral	Large Roman Numeral
5	v	V
6	vi	VI
7	vii	VII
8	viii	VIII
9	ix	IX
10	x	X
11	xi	XI
12	xii	XII
13	xiii	XIII
14	xiv	XIV
15	xv	XV

(continued)

Roman Numerals *(continued)*

Arabic Numeral	Small Roman Numeral	Large Roman Numeral
16	xvi	XVI
17	xvii	XVII
18	xviii	XVIII
19	xix	XIX
20	xx	XX
30	xxx	XXX
40	xl	XL
50	l	L
60	lx	LX
70	lxx	LXX
80	lxxx	LXXX
90	xc	XC

(continued)

Roman Numerals *(continued)*

Arabic Numeral	Small Roman Numeral	Large Roman Numeral
100	C	C
500	D	D
1000	m	M

Notes: _____
_____
_____
_____
_____
_____

Inappropriate Medications for Seniors

Antianxiety Agents

Chlordiazepoxide
Clorazepate
Diazepam
Quazepam
Halazepam
Meprobamate

Antiemetic Agents

Trimethobenzamide

Antidepressant Agents

Amitriptyline
Amitriptyline/chlordiazepoxide
Amitriptyline/perphenazine
Doxepin

Imipramine
Trimipramine

Antihypertensive Agents

Guanabenz
Guanadrel
Guanfacine
Methyldopa

Narcotic Analgesics

Buprenorphine
Butorphanol
Meperidine
Meperidine/promethazine
Pentazocine/naloxone
Propoxyphene
Propoxyphene/acetaminophen
Propoxyphene/aspirin

Sedative–Hypnotic Agents

Amobarbital
Amobarbital/secobarbital
Butabarbital
Flurazepam
Mephobarbital
Phenobarbital
Quazepam
Secobarbital

Sulfonylurea

Chlorpropamide

Notes: _____

Temperature Scales

Celsius (°C) °C = $\frac{(°F-32)}{1.8}$	Fahrenheit (°F) °F = 1.8 × °C + 32
100°	212°
90°	194°
75°	167°
60°	140°
50°	122°
40°	104°
37°	98.6°
36°	96.8°
30°	86°
20°	68°
0°	32°

24-Hour Clock

Note that hours before 12 PM (noon) are pronounced, for example, as "oh one hundred hours" (for 1 AM) or "ten hundred hours" (for 10 AM). Hours after 12 PM (noon) are pronounced, for example, as "fifteen hundred hours" (for 3 PM) or "twenty-two hundred hours" (for 10 PM).

12 AM (Midnight)	2400 or 0000
1 AM	0100
2 AM	0200
3 AM	0300
4 AM	0400
5 AM	0500

(continued)

24-Hour Clock (continued)

6 AM	0600
7 AM	0700
8 AM	0800
9 AM	0900
10 AM	1000
11 AM	1100
12 PM (Noon)	1200
1 PM	1300
2 PM	1400
3 PM	1500
4 PM	1600
5 PM	1700

(continued)

24-Hour Clock (continued)

6 PM	1800
7 PM	1900
8 PM	2000
9 PM	2100
10 PM	2200
11 PM	2300

Notes: _____
_____
_____
_____
_____

Pharmacy Organizations

American Pharmacist Association (APhA) www.pharmacist.com
American Association of Colleges of Pharmacy (AACP) www.aacp.org
American Association of Pharmacy Technicians (AAPT) www.pharmacytechnician.com
American Society of Health-System Pharmacists (ASHP) www.ashp.org
Joint Commission on Accreditation of Healthcare Organizations (JCAHO) www.jcaho.org
National Association of Boards of Pharmacy (NABP) www.nabp.net
National Pharmacy Technician Association (NPTA) www.pharmacytechnician.org
Pharmacy Technician Certification Board (PTCB) www.ptcb.org

State Boards of Pharmacy

Alabama   www.albop.com
Alaska   www.dced.state.ak.us/occ/ppha.htm
Arizona   www.azpharmacy.gov
Arkansas   www.arkansas.gov/asbp
California   www.pharmacy.ca.gov
Colorado   www.dora.state.co.us/pharmacy
Connecticut   www.ct.gov/dcp/site/default.asp
Delaware   www.dpr.delaware.gov
Florida   www.doh.state.fl.us/mqa
Georgia   www.sos.georgia.gov/plb/pharmacy/
Hawaii   http://hawaii.gov/dcca/areas/pvl/boards/pharmacy
Idaho   http://bop.accessidaho.org/
Illinois   www.idfpr.com
Indiana   www.in.gov/pla/pharmacy.htm
Iowa   www.state.ia.us/ibpe
Kansas   www.kansas.gov/pharmacy
Kentucky   www.pharmacy.ky.gov/

Louisiana www.labp.com
Maine www.maine.gov/pfr/professionallicensing/index.shtml
Maryland www.dhmh.state.md.us/pharmacyboard
Massachusetts www.mass.gov/dpl/boards/ph/index.htm
Michigan www.michigan.gov/healthlicense
Minnesota www.phcybrd.state.mn.us
Mississippi www.mbp.state.ms.us/mbop/pharmacy.nsf
Missouri www.pr.mo.gov/pharmacists.asp
Montana http://mt.gov/dli/bsd/license/bsd_boards/ pha_board/board_ page.asp
Nebraska www.hhs.state.ne.us
Nevada www.bop.nv.gov
New Hampshire www.nh.gov/pharmacy
New Jersey www.state.nj.us/lps/ca/boards.html
New Mexico www.rld.state.nm.us/pharmacy/index.html
New York www.op.nysed.gov
North Carolina www.ncbop.org
North Dakota www.nodakpharmacy.com
Ohio www.pharmacy.ohio.gov

STATE BOARDS OF PHARMACY

Oklahoma www.ok.gov/OSBP
Oregon www.pharmacy.state.or.us
Pennsylvania www.dos.state.pa.us/pharm
Rhode Island www.health.ri.gov/hsr/professions/pharmacy.php
South Carolina www.llronline.com/POL/pharmacy
South Dakota www.doh.sd.gov
Tennessee http://health.state.tn.us/Boards/Pharmacy/index.shtml
Texas www.tsbp.state.tx.us
Utah www.dopl.utah.gov
Vermont www.vtprofessionals.org
Virginia www.dhp.state.va.us/pharmacy/default.htm
Washington www.doh.wa.gov
West Virginia www.wvbop.com
Wisconsin www.drl.state.wi.us
Wyoming http://pharmacyboard.state.wy.us

Notes: _____
_____
_____
_____

Index

A

Abbreviations
  diseases, 36–40
  on dosage forms, 10–12
  on drug forms, 16
  drugs, 12–15
  measurements, 17–19
  medical terminology, 19–30
  not accepted by JCAHO, 10
  routes of administration, 31–32
  units of time, 33–35
  vitamins, 35–36
Active immunization
  bacterial vaccine, 2
  toxoid, 2
  viral vaccine, 2
Adrenocortical steroid
  adrenal steroid inhibitor, 6
  glucocorticoid, 6
  mineralocorticoid, 6
Alkylating
  alkyl sulfonate, 1
  nitrogen mustard, 1
  nitrosourea, 1
  triazene, 1
Ambien CR, 45
Amebicide, 1
Amitriptyline, 68
Amitriptyline/chlordiazepoxide, 68
Amitriptyline/perphenazine, 68
Amobarbital, 70
Amobarbital/secobarbital, 70
Anesthetic, 4
Antacid, 7
Antiadrenergic/sympatholytic, 3
  alpha/betaadrenergic blocker, 3
  antiadrenergic
    centrally acting, 3
    peripherally acting, 3
  beta-adrenergic blocker, 3

INDEX

Antianxiety, 4
  benzodiazepine, 4
Antiarrhythmic, 3
Antiasthmatic
  combination, 8
    xanthine, 8
    xanthine-
      sympathomimetic, 8
Antibiotic, 8
Anticholinergic, 7
Anticoagulant
  antithrombin, 2
  heparin, 2
  low molecular weight
    heparins, 2
  selector factor Xa
    inhibitor, 2
  thrombin inhibitor, 2
  warfarin, 2
Anticonvulsant
  benzodiazepine, 4
  hydantoin, 4
  succinimide, 4
  sulfonamide, 4
Antidepressants, 45
  monoamine oxidase
    inhibitor (MAOI), 4
  selective serotonin
    reuptake inhibitor
    (SSRI), 4
  serotonin and
    norepinephrine
    reuptake inhibitor, 4
  tetracyclic
    compound, 4
  tricyclic compound, 4

Antidiabetic
  alpha-glucosidase
    inhibitor, 6
  amylin analog, 6
  antidiabetic
    combination, 6
  biguanide, 6
  incretin mimetic, 6
  insulin, 6
  meglitinide, 6
  sulfonylureas, 6
  thiazolidinedione, 6
Antidiarrheal, 7
Antiemetic/antivertigo
  anticholinergic, 4
  antidopaminergic, 4
  5-HT$_3$ receptor
    antagonist, 4
Antiflatulent, 7
Antifungal, 1
Antihistamine
  alkylamine, 8
  ethanolamine, 8
  phenothiazine, 8, 9
  phthalazinone, 9
  piperazine
    nonselective, 9
    peripherally
      selective, 9
  piperidine
    nonselective, 9
    peripherally
      selective, 9
Antihyperlipidemic
  bile acid
    sequestrant, 3

fibric acid derivative, 3
HMG-CoA reductase
inhibitor, 3
Anti-infective, 6
drugs, 1
Anti-inflammatory, 6
nonsteroidal selective
COX-2 inhibitor, 4
Antimalarial, 1
Antimetabolites, 1
Antimitotic
taxoid, 1
vinca alkaloid, 1
Antineoplastic
drugs, 1–2
Antiparkinson
anticholinergic, 4
dopaminergic, 5
Antiplatelet
aggregation
inhibitor, 2
antiplatelet
combination, 2
glycoprotein IIb/IIIa, 3
Antipsychotic
benzisoxazole
derivative, 5
dibenzapine
derivative, 5
dihydroindolone
derivative, 5
phenothiazine
derivative, 5
phenylbutylpiperadine
derivative, 5

quinolinone
derivative, 5
thioxanthene
derivative, 5
Antiretroviral
fusion inhibitor, 1
nucleoside analog
reverse
transcriptase, 1
nucleoside reverse
transcriptase, 1
protease inhibitor, 1
Antirheumatic, 2
gold compounds, 2
Antiseptic, 8
Antitoxin/antivenin, 2
Antiviral, 1
Artificial tear, 8
Astringent, 6
Avonex, 45

B

Bacitracin, 1
Benzodiazepine, 4
Body surface area,
41–42
Bronchodilator
anticholinergic, 9
diluents, 9
sympathomimetic, 9
xanthine derivative, 9
Buprenorphine, 69
Butabarbital, 70
Butorphanol, 69

INDEX

C

Calcium channel blocker, 3
Cephalosporine, 1
Chemotherapy regimen, 2
Chlordiazepoxide, 68
Chlorpropamide, 70
Cleanser, 6
Clorazepate, 68
CNS stimulant
   amphetamine, 5
   analeptic, 5
   anorexiant, 5
Coagulant heparin antagonist, 3
Corticosteroid, 8

D

Detoxification
   antidote, 6
   chelating agent, 6
Diaper rash, 6
Diazepam, 68
Diseases, 36–40
Diuretic
   carbonic anhydrase inhibitor, 7
   diuretic combination, 7
   loop diuretic, 7
   nonprescription diuretic, 7
   osmotic diuretic, 7
   potassium-sparing diuretic, 7
   thiazide and related diuretic, 7
DNA, 2
Dosage calculations, 41–42
Dosage forms, 10–12
Doxepin, 68
Drug forms, 16
Drugs, 12–15

E

Elidel, 45
Emsam, 45
Enzyme, 2
Epzicom, 45
Expectorant, 9
EXUBERA, 45

F

Fluoroquinolone, 1
Flurazepam, 70
Foradil, 45
Forteo, 45

G

Gastrointestinal, 7
   anticholinergics/antispasmodic, 7
Gastrointestinal stimulant, 8
Genitourinary irrigant
   hexitol, 7
   neomycin and polymyxin B, 7

Glaucoma, 8
  alpha-adrenergic agonist, 8
  alpha-adrenergic antagonist, 8
  beta-adrenergic blocking agent, 8
  carbonic anhydrase, 8
  miotics
    cholinesterase, 8
    direct acting, 8
  prostaglandin agonist, 8
  sympathomimetic, 8
Guanabenz, 69
Guanadrel, 69
Guanfacine, 69

H

Halazepam, 68
Histamine H_2 antagonist, 8
Hormone
  androgen, 2
  antiandrogen, 2
  antiestrogen, 2
  aromatase, 2
  progestin, 2
Hospital chart information
  maximum daily dosages, 47
  order requirements, 46

I

Imipramine, 69
Immune globulin, 2
Immunologic
  immunomodulator, 2
  immunostimulant, 2
  immunosuppressive, 2
Impotence
  phosphodiesterase type 5 inhibitor, 7
INFERGEN, 45
Isotretinoin, 45
IV flow rate, 43
  shortcut, 43
IV infusion time, 43
IV volume, 43–44

J

JCAHO, abbreviations not accepted by, 10

K

Keratolytic, 6

L

Lariam, 45
Laxative, 8
Lincosamide, 1
Lindane, 45
Lipase inhibitor, 8
Lipopeptide, 1
Lotronex, 45

INDEX

M

Macrolide, 1
Measurements, 17–19
Medical terminologies, 19–30
Meperidine, 69
Meperidine/promethazine, 69
Mephobarbital, 70
Meprobamate, 68
Methenamine, 1
Methyldopa, 69
Metric conversions
  household, 48–49
Mifeprex, 45
Migraine
  ergotamine derivative, 5
  serotonin 5-HT$_1$ receptor agonist, 5
Muscle relaxant
  depolarizing neuromuscular blocker, 5
  nondepolarizing neuromuscular blocker, 5
  skeletal
    centrally acting, 5
    direct acting, 5

N

Nasal decongestant
  arylalkylamine, 9
  imidazoline, 9
Nolvadex, 45
Nonnarcotic antitussive, 9
Nonsteroidal antiinflammatory drugs, 45

O

Ocular lubricant, 8
Opioid analgesic, 5
Otic preparation, 8

P

Pacerone, 45
PEGASYS, 45
PEG-Intron, 45
Pentazocine/naloxone, 69
Percent of a quantity, 44
Pharmacy organizations, 75
Phenobarbital, 70
Photochemotherapy, 6
Pigment agent, 6
Plasma expanders, 3
  dextran adjunct, 3
  plasma protein fractions, 3
Prefixes, 50–53
Pregnancy categories, 57–58
Professional titles, 59–62
Propoxyphene, 69
Propoxyphene/acetaminophen, 69
Propoxyphene/aspirin, 69

Prostaglandin, 8
Proteasome inhibitor, 2
Protein-tyrosine kinase inhibitor, 2
Proton pump inhibitor, 8
Protopic, 45
Punctal plug, 8

Q

Quazepam, 68, 70
Quinolone, 1

R

Ratio–proportion method (two equivalent ratios), 44
REBETRON, 45
Rebif, 45
Renin angiotensin system antagonist, 3
   angiotensinconverting enzyme inhibitor, 3
   angiotensin II receptor antagonist, 3
   selective aldosterone receptor antagonist, 3
Respiratory inhalant corticosteroid, 9
   intranasal steroid, 9
   mast cell stabilizer, 9
   mucolytic, 9
   respiratory gases, 9
Retail hardcopy information, 63
Retinoid, 6
Retinoide, 2
REVLIMID, 45
Rexinoid, 2
Ribavirin, 45
Roferon A, 45
Roman numerals, 64–67
Routes of administration, 31–32

S

Salicylate, 5
Scabicide/pediculicide, 6
Secobarbital, 70
Sedative and hypnotic, 5
   barbiturate
      intermediate acting, 5
      long acting, 5
      short acting, 5
   nonbarbiturate, 5
Serevent DISKUS, 45
Sex hormone
   anabolic steroid, 6
   androgen, 6
      hormone inhibitor, 6
   contraceptive hormone, 6
   estrogen, 6
      and progestin combination, 6
   ovulation stimulant, 6
   progestin, 7

INDEX

Sex hormone *(conti.)*
    selective estrogen receptor modulator, 7
Smoking deterrent, 5
Soriatane, 45
State Boards of Pharmacy, 76–78
Strattera, 45
Sucralfate, 8
Suffixes, 54–56
Sulfonamide, 1, 8
Sunscreen, 6
Surgical adjunct, 8
SYMLIN, 45

T

Temperature scales, 71
Tetracycline, 1
Thrombolytic
    enzymes, 3
    recombinant human activated protein C, 3
Thrombolytic
    tissue plasminogen activators, 3
Thyroid
    antithyroid agent, 7
    hormone, 7
Tracleer, 45
Trimethobenzamide, 68
Trimipramine, 69
24-hour clock, 72–74
Tysabri, 45

U

Units of time, 33–35
Upper respiratory combination, 9
Urinary acidifier
    acid phosphate, 7
    ascorbic acid, 7
Urinary alkalinizer, 7

V

Vaginal preparation, 7
    anti-infective miscellaneous, 7
    vaginal antifungal agent, 7
Vasodilator
    endothelin receptor antagonist, 4
    human B-type natriuretic peptide, 4
    nitrate, 4
    peripheral vasodilator, 4
    prostacyclin analog, 4
Viramune, 45
Vitamins, 35–36

W

Wound healing agent, 6

X

Xyrem, 45

Z

Ziagen, 45